TEACHING AND LEARNING MATHEMATICS THE BEST WAYS:

A Guide for Parents, Teachers and Students

DAVID KUNDA MVULA

Copyright © David Kunda Mvula, 2020

The moral rights of the author have been asserted.

First edition published: 2020

All rights reserved. No part of this publication may be reproduced or transmitted, in any form or by any means now known or to be invented, electronic or mechanical, including photocopying, recording or by any information storage or retrieval system, without the prior permission in writing of the author, or as expressly permitted by law or by licence, with the exception of brief excerpts in magazines, articles, reviews, etc.

This work must not be circulated in any other form except that prescribed by the author and the same condition must be imposed on any acquirer.

ISBN: 9798734626122

Editor: Sydney Matakala Kasanda for Libros Editing and Publishing
sydneyk@librosediting.com

Author Contact Details:

Mobile Number: +260 97 5339212

Email: mathsthebestway@gmail.com

Facebook: Learn/Teach Mathematics The Best Ways

DEDICATION

This book is a special dedication to all the students of mathematics, to all lovers of this subject, to the students who hate mathematics and see it as a barrier to their career paths, to all the parents and guardians who aim to help and nurture their children so that they can reach greater heights in learning mathematics the best ways, and to all the teachers of mathematics.

You are the reason this book was written.

ACKNOWLEDGEMENTS

In my efforts to come up with this book, a lot of people contributed so much in different ways. Listing them all would make a read for another day. However, I would be remiss if I did not make mention of particular persons to whom the publication of this book is greatly owed.

I am forever indebted to my parents, Mr. and Mrs. Mvula, for their immeasurable support and love. They sacrificed their happiness just so I could have the second greatest gift – knowledge and wisdom. May God bless them even more!

My gratitude goes to my brothers and sisters for their generosity: the Mvulas, Gomas, Fundas, Mukondes, Mumbas, Sikazwes, Nyimbilis, Chondokas, Chilufyas, Kalubas, Chaushis, and Yambayambas, among others. To you all, I am grateful.

To all my friends, your support is something unmerited. My friend Sinjani Innocent, you helped me to develop the Mvula`s theorem, which in turn gave birth to the ideas in this book. I am grateful. To Mwananshiku Fredrick, a friend who introduced me to Kamuchanga Library where the ideas in this book started developing, I am grateful. To Twenda Elijar Greg, you are everything I would ask for in a friend. Your love and support is incomparable. To Mr. Mwansa Sailas, who wholeheartedly offered his personal computer that I used to type the first manuscript, I cannot thank you enough.

To comrades David Kashiki and Balewa Zyuulu, your passionate works towards supporting others and authoring re-ignited my dream to finish my book. I am grateful.

To all my former teachers for your mentorship and support throughout my life, I am forever thankful.

In my writing and re-writing of this book's manuscript, my family – my wife and children – were the most affected as I spent most times, days and nights sometimes away from home. To the most loving and supportive woman I have ever known, Violet Chilufya Mvula, thank you. To my sons, Gomezyani Frank Mvula and Aninjifya David Mvula, your love pushes me to greater heights day in and day out. To have you is a blessing, I am grateful.

My sincere appreciation also goes to Mr. Sydney Kasanda, firstly for understanding the importance of the content of this book, and secondly for the editing services he provided.

Lastly, but not the least, I give thanks to God – who is the greatest teacher ever. And in the scriptures of *Philippians 4:8* He guides us to always, always keep considering all things of great value and importance. Therefore, in all things, to God Be the Glory!

MESSAGES OF ACCLAIM

"I can't hesitate to tell you that this is the best book I have ever come across on how one can learn mathematics the best way; everything written in it is top notch, and I am certain it will encourage many to develop intense interest in mathematics." – **Nthandi Njelezye, Copperbelt University Meteorology Engineering Student**

"These are the ideas we want… very explicit and to the point…" – **Njobvu Paul, Mufulira College of Education Senior Lecturer of Mathematics**

"When I just saw the title, I said 'wow!' This is a good one. I have scanned through the book, and this is a good piece of work for a start." – **Koji S.K.M, Head of Section Mathematics Department, Mufulira College of Education**

"The expert in any field was once a beginner, hence to consider becoming exceptional in mathematics as a student, parent or as a teacher, I highly recommend this well-articulated and simplified gift in form of a book by David Mvula." – **Fred Kabaso, BEng. Environmental Engineer from the Copperbelt University, Environmental Consultant, Pastor, Entrepreneur and Youth Leader**

*"I refused to edit the manuscript because I didn't think I was qualified to do it. After all, I performed poorly in mathematics that it doesn't even feature on my 1983 high school certificate. Nonetheless, I took

a chance and accepted to look at it knowing I would not do it. Once I read the script, I knew I wanted to edit it. Since then, I have gone back to school to re-sit my Grade 12 mathematics. It is never too late." – **Sydney Kasanda, Editor at Libros Editing and Publishing**

"There is so much substance in here." – **Humphrey Bwalya, Secondary school teacher of Mathematics and Physics, Northrise Combined School of Excellence**

"I like it when you say something like 'mathematics is attached to every aspect of our lives'. You are really a great writer!" – **Funda Lewis, Kansanshi Mines Plc.**

"This book is an eye-opener and a guide not only in the teaching and learning of mathematics the best ways, but in education and life in general. Thank you David" – **Felix Mwansa, Author/Motivational & Inspirational Speaker/Lecturer/Chartered Accountant**

FOREWORD

'I should have known this earlier,' I thought to myself as I read this book. We surely cannot be disciplined at all times. To achieve discipline and eventual success, you have to lose to gain sometimes, and vice versa. To lose can also mean subtraction of worthless things from your life, and gain as the addition of things that are worthwhile. Addition and subtraction in every sense is mathematics. We cannot live without mathematics in any way – it is impossible to have a normal life without its application. Clearly, mathematics is indispensable - any addition and subtraction of any sort is mathematics.

This book reveals the minute secrets we have been missing, and the tips given here are simple and effective. David Kunda Mvula brings to light that teaching and learning mathematics is not a one man-army: teachers, parents, and students must work together to achieve their intended goal. Continuous improvement through knowledge building programmes sharpens their skills and approaches.

Students' weaknesses and strengths must be understood and attended to, and help them to weave the concepts into their daily activities beyond theory and exams. They should also have a positive attitude toward this seemingly "hard" subject and appreciate the connectivity between previously learnt concepts and the current. Undoubtedly, it is the lack of interest that causes most students to perform poorly rather than the hardness of the subject. Read and you will never view teaching and learning of mathematics the same way.

Abigail Sandi,

Writer and Mentor, Environmentalist (Bachelor of Environmental Education)

TABLE OF CONTENT

PREFACE..13

Chapter One

1. **INTRODUCTION**...17

   1.1 How Do You Look at the Teaching and learning of Mathematics..20

   1.2 .The Right View at Teaching and Learning of Mathematics..20

Chapter Two

2. **WHAT IS MATHEMATICS?**.....................................23

   2.1 When, Where and How Did Mathematics Begin?...............26

Chapter Three

3. **IS MATHEMATICS IMPORTANT? WHY SHOULD IT BE TAUGHT TO ALL?**...29

   3.1 What's Your View About Mathematics as a Subject?.........29

   3.2 Is Mathematics Really Important?..........................30

Chapter Four

4. **HOW BEST SHOULD MATHEMATICS BE TAUGHT?**.........37

   4.1 How Should Mathematics Be Learned?.....................45

   4.2 Getting Involved...47

Chapter Five

5. **STUDENTS AND PARENTS ROLES IN THE TEACHING AND LEARNING OF MATHEMATICS**.................................53

   5.1 Teachers' and Students' Roles 53

5.2 Parents' Roles ………………………..…………………..57

5.3 The Role of Cooperating partners ………..……………..61

Chapter Six

6. HOW TO SUCCEED AT MATHEMATICS EXAMINATIONS…63

6.1 Ten Points on the Best Way to Prepare For Examinations………………………………………...…………64

CONCLUSION……………………………………….…..69

PREFACE

'Teaching and Learning Mathematics the Best Ways: *A Guide for Students, Teachers, and Parents*' is about mathematics education. It is a guide in the sense that it provides readers with thoughtful and proven best ideas, and approaches gained through many years of teaching and learning of mathematics, and also through interactions with the great minds of students, teachers, and parents. This book also shares ideas and experiences of parents and teachers concerning teaching and learning mathematics the best ways.

I decided to write this book upon realizing the true value of mathematics to human kind. Its value, which I believe is relevant to the survival of humans, must be shared to all. Secondly, this book was written as a response to the wrong prejudices and stories about mathematics education. As human beings, we all use mathematics whether consciously or subconsciously. Therefore, by understanding how to teach and learn mathematics the best ways as guided herein – from students', teachers' and parents' perspectives, readers are assured of proven success in teaching and learning of mathematics.

This book provides direction and encouragement – firstly to those who have been engulfed in fear and have declared mathematics to be a difficult and boys' subject, but are determined to improve their performance in it. Secondly, it provides motivation to the lovers of mathematics to further their love for mathematics. Furthermore, it is also a guide to the teachers as well as the parents on how best they can assist their children to do better in this subject, whether or not they themselves had impressive performances in mathematics in their former years. To all the parents whose children have declared that the study of mathematics as a subject is too hard, this is a must read.

It is said, the roots of education are bitter, but its fruits are sweeter. The concepts and ideas contained in this book might have been written or talked about by others in many ways, and may imply that it is simply an addition to an already existing body of knowledge of mathematics. However, this book is the latest of its kind. It interacts with you as a reader as though you were talking to the author, taking into account the latest best approaches to teaching and learning of mathematics.

Employing the ideas herein will call for total willingness on the part of students, teachers, and parents to learn new ideas and strategies, as well as to unlearn some of the already learnt ones. This might not be easy, but it is rather simple. There will be bitter times, but the outcome is sweet. Man and mathematics as a subject are inseparable. The nature of man and the activities he performs in his environment makes him a mathematician by nature. The use of man and the related pronouns in no way excludes the woman in this context. In every second of our lives as human beings, we are subconsciously practicing and solving mathematical problems.

Mastering of mathematical ideas, learning, and teaching strategies makes man's life great. To achieve this greatness, man cannot work in isolation. Students, teachers, parents, and other stake holders need to all take up their roles in the teaching and learning of mathematics more seriously. This must be done joyously, harmoniously, and lovingly, and mathematics will smile back at mankind.

David Kunda Mvula

CHAPTER ONE

1. INTRODUCTION

A lot has been said and written about mathematics – what it is as a subject, how it begun, and how it should be learnt or taught. To begin with, mathematics is simple. It is interesting and valuable.

Is mathematics *simple*? Perhaps this might be the question lingering on your mind right now. To say it simply – YES, it is! How simple is the teaching and learning of mathematics? Actually, this plus more is what this book unveils in the coming chapters. Unhappily, I should admit though that the opposite is what some people have perceived it to be – that mathematics is too hard or that it can only be effectively learned by boys. So, I encourage you to move with me and digest the ideas in this book as it offers readers a unique perspective to mathematics education.

Mathematics is a living subject. It has always been there throughout man's existence. It is eternal and a combination of both inventions and discoveries. Mathematics as a subject is a part to man's life, hence is real just as man is. A child or a student in early grades is like a seed. Its future is not yet known, and for that, proper cultivation is inevitable. The nurturing of a child first comes from the families in homes or the environments in which, like seeds, they get placed. Like gardeners, teachers – as parents in schools, do not know exactly what each child will become. The gardener does not know what will become of the seed. However, teachers, unlike gardeners, do not receive seed packets or instruction manuals.

For this reason, teachers as well as parents must understand that provision of an environment which is conducive for learning is of great importance in ensuring the full development of a child. A seed will develop to its full potential in an environment that is full of appropriate nutrients; otherwise it will grow yellowish, stagnate, and eventually die. Janisse Ray in her 2012 book, 'The Seed Underground: *A Growing Revolution to Save Food*' says, "A seed makes itself…, but it needs help. Sometimes, it needs a moth or a wasp or a gust of wind. Sometimes, it needs a farm and it needs a farmer. It needs a garden and a gardener. It needs you." (*hhtps://yesmagazine.org/issues, 05/04/2013*)

In the same sense, students need a suitable environment for them to maximize their full potential. They need support from teachers and parents, and they need you. Interaction is one of man's characteristics, and through it, human beings interact with and within their environment. This brings about change and growth of the mind. The development of the mind or mental growth in humans comes about as a result of how they react to their environmental factors. Environmental factors always have a great impact on learning. On a broader scale, talking of the environment implies human habitation in a full sense. It does not only refer to our physical environment such as our homes, schools, and locality, but also the people around us – **the education we receive, the social customs, and values and traditions**. Also important to note is the fact that environmental factors do not shape individuals, but how each individual reacts to them makes them what they become in life.

But why should we talk about learning or human mental development in relation to the environment? Of what importance does it have to the teaching and learning of mathematics? It is for the simple reason that the exposure to the environment is of a greater impact on the child's growth, and also on the mathematical (academic) performance of individuals in schools and life in general. In fact, a number of studies show that the environment has a remarkable influence on the learning of individuals.

I decided to talk about the teaching and learning of mathematics in connection to the environment, keeping in mind that there are some environmental factors at play. Whilst learning as a process starts at birth, the influence of the environment starts as early as the conception of a child in the mother's womb. Also, while the hereditary factors might also be at play, it is the experiences gotten from the environment that play a major role in the development of mathematical abilities in a learner. The learning of mathematics begins as early as at a child's conception stage and ends at death, which is man's last experience. Early mathematical experiences learned both at school and home go a long way in the students' lives.

1.1 How do you look at the Teaching and learning of Mathematics?

To 'some' students, mathematics as a subject is sometimes considered as a barrier to certain career paths, while some passionately detest mathematics as a compulsory subject and look at it as an enemy to their success. Others wish as much as possible to avoid anything to do with mathematics. To others, mathematics is a source of pleasure and a key to problem solving skills. How was the situation during the inception of this subject? Let us look back a little.

1.2 The Best View at Teaching and Learning of Mathematics

At its inception, mathematics was largely known to be a boy's subject. Unlike with other subjects, it was considered to be suitable only for boys and a few girls, but mostly those who showed the special ability to deal with mathematical problems. This is evident through considering the fact that many prominent ancient mathematicians were male, looking at mathematicians such as Pythagoras, Euclid, Carl Guass, Descartes, Isaac Newton, Archimedes, among others. In those days, girls and other students with shallow levels of intelligence were considered to be unqualified to study mathematics.

However, that in itself does not eliminate girls from the circles of mathematics. As a matter of fact, there are some ancient and modern female mathematicians worth noticing and emulating. Hypatia (350-

415 AD) of Egypt is one of them, and Katherine Johnson (1918-2020), a female black mathematician and her historical contributions to NASA safe space travel, *https://www.dreambox.com/blog/famous-female-mathematicians, 18/05/2020*.

Over the years, mathematics as a subject has continued to grow. Now more than ever, we are seeing many mathematical discoveries and inventions being made. At least, people in this technological age have come to realize the importance of mathematics to everyone's life. The study of mathematics is NOT as difficult as it was perceived to be during its inception. Everyone now must understand that mathematics is not a 'boys' subject only'. It can be studied by both sexes. One key thing is the love for the subject. As earlier stated, there are a lot of examples we can give today of girls who have excelled at the learning of mathematics. For instance, since its inception, the Copperbelt University had never produced a graduate with a distinction in the field of civil engineering, a field which is more mathematical in nature. However, in the year 2019, Marble Musabi, a female student, graduated with a distinction – an indication that girls can excel in mathematics just as well as men do, if not better. We shall look at the love for mathematics later.

The increase in the numbers of both male and female mathematics students worldwide is an indication that the trend of realizing that mathematics is a subject for all has spread to every corner of the earth. For example, during my diploma studies at college where I studied mathematics as a single major course, our class had about seventy-five students out of which at least forty were female students. The myths and some people's prejudice about mathematics as the subject for boys only and intelligent ones are still alive. It remains a great threat to the teaching and learning of the subject. As a parent, student, and teacher of mathematics, this is a challenge I face constantly. I strongly oppose these myths and prejudices. Not until us as teachers, students of mathematics, and parents work together on this, all our efforts will be in a drop in the bucket. I mean, how on earth will we succeed while the old myths are still standing tall? There is a dire need for all of us to work on breaking down the prejudices and myths that have engulfed the majority about mathematics education.

Mathematics is in everything about life. It is in the way man organizes his life activities. Its application is the start to all man's activities and the end to it, and its mastery defines the difference between success and failure. With all of man's potential, success can only be achieved with the mastering of mathematical concepts implying that not only does this book guide about the best way of teaching and learning of mathematics, but it also guides **teachers, students, and parents the right perspectives towards the teaching and learning of mathematics.** It is about how best they must look at the teaching and learning of mathematics as the insiders of the learning process rather than the outsiders, how they should relate during the pre-tertiary level, and also the best approaches and strategies to apply.

CHAPTER TWO

2. WHAT IS MATHEMATICS?

It was at secondary education level when I was first introduced to subjects like chemistry, geography, religious education, and others. I still remember that in all of these subjects, my teachers started by defining the subjects and the subjects' content they were teaching us. This helped in giving us (the students) a clear picture of what each subject was all about. However, this was not, and is still rarely the case with many teachers of mathematics. In schools, mathematics as a subject and its subject content is rarely defined. But, are the definitions in mathematics really important? Yes, there are! In fact, when a definition is not given, it is from such instances that some students begin to lose their way to learn and understand mathematical concepts. They start to imagine their own definitions, and this leads them into misconceptions and loss of interest in the subject. Definitions are very important in that they enable students to have a clear and summarized view of the whole subject. This is essential especially at the introduction of every mathematical concept.

"Most concepts in everyday life like house, orange, cat, etc. is acquired without any involvement of definitions. On the other hand, some concepts even everyday life concepts, might be introduced by definitions. The word "forest" might be introduced to a child by saying "many, many trees together". It is definitions like this that help to form a concept image.

https://www.researchgate.net/publication/227282219_The_Role_of_Definitions_in_theTeaching_and_Learning_of_Mathematics, 22/05/20.
Definitions act as scaffolds. Without them, it is almost impossible that a tall building can be constructed. Similarly, without definitions, it is almost impossible that students can fully grasp mathematical concepts. Many educationist and mathematicians have across the globe tried to define what mathematics is. Benjamin Peirce (1809-1880) termed mathematics as, "the science that draws necessary conclusions." *htps://quotes.yourdictionary.com/author/quote/609060*)

I look at mathematics as a line of thoughts and activities that lead to solid conclusions. Any activity which is conclusively solid and provable is mathematics – for Instance, the concept of Pythagoras theorem shown in Figure 1.

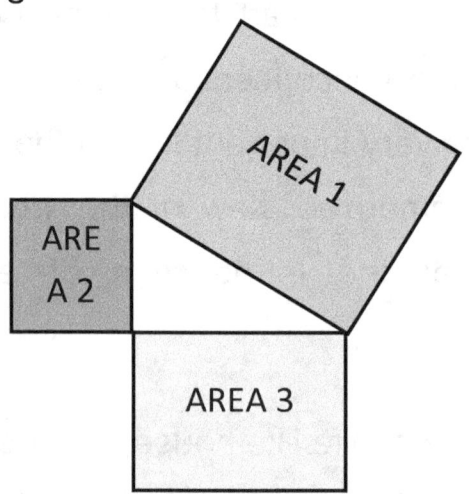

Figure 1

In the diagram (*Figure 1*), AREA1² = AREA2² + AREA3². These are solid conclusions and they can be proven considering that the shapes in figure 1 for Area1, Area2 and Area3 are all squares. Mathematics can also be said to be the reasoning of some truth in a certain context, situation or pattern that is provable. It is the study and proper use of symbols, signs, and numbers in making up solid conclusions.

When we say that Frank is older than Gift, and that Gift is older than Aninjifya, then Frank is older than both Gift and Aninjifya. This kind of reasoning gives a solid conclusion, and is mathematical.

Picture 1: Frank (left), Aninjifya (center) and Gift (right), 11/08/2019

Another example of mathematical reasoning is on the perimeter of a rectangle. It is calculated by finding the sum of two lengths and two breadths. This is the reasoning of some truth because a rectangle has four sides – two lengths and two breadths, and the conclusion is that P (perimeter) = 2 (length + breadth).

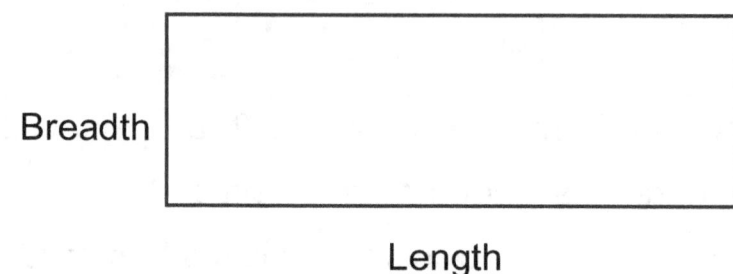

Figure 2   Length

2.1 When, Where and How Did Mathematics Begin?

Mathematics has no specific place of origin. Considering its many definitions gives great evidence that it begun from different parts of the world and at different times. Mathematics has been defined by people from all across the world, giving an indication that it begun at different times and places.

As a subject, mathematics came about because of its application value. It came about because of its usefulness in different fields such as science, engineering, medicine, and economics, among others. Mathematics, which has now come to be known as the global subject, begun as early as 6BC with the Pythagoreans, who coined the term 'mathematics' from the ancient Greek (*Mathema*) meaning science, knowledge, and learning https://wikipedia.org.

At its inception, mathematics was only in a few locations. The most ancient mathematical writings are known as the Babylonians (*Plimpton 322*) discovered as early as 1900BC. Then there is the Egyptian mathematics discovered around 2000–1800BC. The expansion of mathematics as a subject came as a result of the great works of the Greeks, who mostly did the refined work on deductive reasoning and mathematical proofs. In addition, there is the Chinese mathematics, which includes the place value system, and the Hindu-Arabic numeral systems. These played a greater role on the beginning of mathematics.
(*Howard E.,1990*).

As early as 3000BC, Egyptians where familiar with geometry such as circles, ellipses, and the Pythagoras triples in their designs (*Kulbur S.,1995*). They used the mathematical knowledge of geometry in their civilization which they applied in equal sharing of land to its people. The Babylonians are well known for the discovery and use of sexagesimal (base 60) numeral system. It is from this that the modem day use of 60 seconds, 60 minutes in an hour, and 360 (60x60) degrees was derived from.

Among the Greek great works, we find the Pythagoras theorem, the Thales theorem, and in fact, Thales of Miletus (546BC-624BC) is hailed as the first mathematician to have used the deductive reasoning applied in geometry, and as the individual to whom mathematics discovery has been attributed (*wikipedia.com*). It is

Pythagoras who established the Pythagorean School whose doctrine was that mathematics ruled the universe and the motto 'All is number'. Mathematics uses symbols. It is these symbols we call numbers that simplify life and all activities one can think of.

In defining mathematics, its development, and its origin, there are many names of contributors or mathematicians that can be called to list – ancient and modern, male and female – but since its development is perpetual, many discoveries are still being made, and shall continue to be made. Therefore, what is important is not to know all the names of those people, but rather to understand that mathematics is the subject that can be studied by everybody from everywhere as long as there is interest and love for the subject.

CHAPTER THREE

3. IS MATHEMATICS IMPORTANT? WHY SHOULD IT BE TAUGHT TO ALL?

A lot has been said about mathematics as either being a compulsory or an optional subject; also about its importance and the reasons to why it should be taught and learnt by all.

3.1 What's Your View about Mathematics as a Subject?

"Mathematics is not important because it is a subject used just to torment weak students' minds by those who understand it well," Gloria (not real name) one of my former students once lamented. Do you share a similar view that mathematics is only a tool used to torture the weak students' minds? Well, students' experiences offer different opinions and reasons for such arguments, and should be respected though some maybe incorrect – yes, incorrect!

First of all, I must mention that in as much as the teachers, parents, and other interested parties stress out about the importance of teaching and learning mathematics by all, it should be understood that it is not easy for the students to easily see 'that' importance in the theoretical classrooms in which mathematics students are usually confined. For instance, concepts such as the trigonometric functions and geometry cannot easily be seen in the real world. Therefore, unless we consider the importance of teaching and learning mathematics from its utility point of view, all efforts will be in vain.

Again, one thing is clear though, whether or not students enter a field that is more dependent on the use of mathematical problem solving skills and knowledge. It is a fact that they will in their real life situations have a need to apply approaches that are mathematical in nature. Man cannot be separated from the circles of mathematics.

For a long time now, people have been debating whether or not mathematics should be taken as a compulsory subject in schools at pre-tertiary level. In the year 2009, and for the first time in Zambia's educational history, it was recommended that any student was to be considered qualified to the senior secondary level (Grades 10-12) as long as he/she met the six (6) subject pass either with or without mathematics. This meant that mathematics was no longer considered compulsory or rather as a very important part of learning for every student. Rationally, it was made optional. ('Every subject is to be treated with the same importance' *Ministry of Education, Educating our future policy, 2006*).

3.2 Is Mathematics Really Important?

Is the mathematics learnt at primary and junior levels enough that students should not be taught mathematics at senior level? Should it be compulsory or optional in schools? Remember, "All is number". Should what makes all man's activities be optional? Dennis Sedney

Nawaseb of University of Namibia (2012) asked the same question online as she was conducting her university research project, and I picked the following from the comments box:

"Mathematics should be taught in all schools and made compulsory at pre-tertiary levels. This is so because it gives the necessary foundation required in other subjects. The structure of mathematics is that it is all embracing and serves as the desired tool used for human development" (Ogunkule Abiodun, 2012).

"To survive socially and economically in today's society, you must have math…" (Zinn Pamela, 2012).

Additionally, in the mathematics count, a report from the committee of inquiry into the teaching of mathematics in British schools under the chairmanship of W. H. Cockcroft, the question "Why teach mathematics?" was answered as; *"It would be very difficult – perhaps impossible to live a normal life in very many parts of the world in the Twentieth Century without making use of mathematics of some kind."*

In answering more to this important question, let us consider the following scenarios:

Scenario 1
We have all served meals before or at least we have had a chance to be served by someone. When preparing any meal, normally a person asks oneself – what is the occasion? What specifically should I

prepare? What is the quantity of the ingredients needed? How many people are to be served from the meal? And what kind and size of utensils are needed?

Scenario 2

When making a personal or house budget, one also needs to look at how much income is available. He/she should consider how much is to be spent, and ensure that the budget sustains all the needs for the specific period. Who is not able to do all that anyway?

Perhaps that is what you are asking. Well, the significance of mathematics starts from as simple as these very examples. What are the two above scenarios about? What do they involve? Are these activities mathematical? When either budgeting or preparing for meals, you can agree with me that there is consideration of the sum of money available, the number of people included, and the utensils required. There is also the division of what is available for the other needs. Do these activities not summarize man's life? Is it not the addition, subtraction, multiplication, and division that make up the four order operations of mathematics? Is there an Individual who does not need such knowledge? Should that be optional to man's life? I guess your answer is as good as mine.

Further, an athlete who gains a lot of weight will be asked to lose some kilos because being overweight could negatively affect his/her performance. That will be subtraction in weight. Is that not mathematics? Is it not important? Imagine the consequences of one defaulting on a mortgage payment if for example one wrongly **calculated** the payment dates? The consequences can be horrifying. That is mathematics.

Try also to think of the havoc that would wreak society if for instance the pharmacists or medical doctors did not understand the mathematical measurements of the content of the dosage to be administered to their patients. Think of the hazards that would be created by an engineer constructing a tower, a bridge or a house, but lacks the mathematical skills to do accurate measurements. Think of how needy a nation would be with a government or citizenry whose knowledge is devoid of mathematics. These, among other daily man's routines, solidify the fact that mathematics is important. It forms all the activities of man's life, and should be compulsory at least at the pre-tertiary level.

Mathematics is not just important to man's life, but its usefulness and precise conclusions are very astonishing. This is why it is referred to be the greatest art of all arts, and the science of all sciences. It is the backbone to most, if not all great arts and sciences in developed countries around the world. Consider the contributions of mathematics to the world's civilization. As earlier alluded to, the Egyptians where able to use mathematics to ease the sharing of its land to the people and for construction. Egypt was largely civilized because the Egyptians where good in geometric mathematics. Even today, the ancient pyramids, which are the work of the Egyptians, are a marvel and a testament of the great application value of mathematics.

Countries such as Japan, China, Great Britain, France, Germany, and Singapore, among others, are well developed today because they have all put mathematics and science at the core of everything. Just think of how difficult and costly it would have been to move from one country to the other without the knowledge that gave birth to the technology of telephones and the internet, also the knowledge of the world map. Imagine also what life would be without the invention and knowledge on the proper use of mathematical instruments such as computers, direction compasses or calculators. But thanks to Martin Cooper for the mobile phones, and the Chinese Han Dynasty for the invention of mathematical instruments in circa 1973 and 206BC respectively. Galileo Galilei (1564-1641), the man who moved across the earth invented the mathematical Instruments, and came up with the estimates of the earth's circumference (40,000km), and later invented the telescope. The world is a global village – thanks to mathematics.

Lastly, check your environment. Check where you are seated, the structures around you. Are you not able to see mathematics around you? How well can the importance of mathematics be explained than this? Indeed, the vitality of this great subject is there for all to appreciate. Mathematics is useful in all fields or subject areas; in religion, health, economics, and education. Christians give 10% (the calculation of the salaries' 10% is mathematics) of their income as tithe. Geographers and pilots use mathematics when finding locations on the world map. Mathematics is important because of its usefulness in many fields – for communication purposes, enabling logical thinking, for pleasure, and problem solving skills.

Mathematics is important and unique. For instance, the use of vehicle number plates (mathematical symbols) and mobile phone numbers have proved to be an effective means of communication. There are many, yet distinct. In fact, if mathematical ideas were to disappear, the whole world would be in confusion, perhaps even the earth would begin to disappear too. It is mathematics whose harmonious ideas, processes, and activities that holds the universe together. It is the mathematical activities which enable all harmonious, effective, and fundamental life processes on earth. This is why every student deserves a compulsory study of it. Hence, is mathematics important? YES, it is Important, and it should be learnt in schools by all as a compulsory subject.

CHAPTER FOUR

4. HOW BEST SHOULD MATHEMATICS BE TAUGHT?

Since ancient times, teachers have always played a very important role in the teaching and learning processes of mathematics. Back then, the betterment of every learner's performance in mathematics was dependent mostly on the teachers' effectiveness in fulfilling their role in school. Those days, teachers were regarded as the "know it all", and pumping knowledge into the students' minds like in empty buckets was their way of teaching. Was this the best way of teaching and learning mathematics?

It is said that you cannot pour grain in a sack unless the sack is open to receive it. Similarly, your subconscious cannot pour ideas into your mind unless your mind is receptive (*Estelle H. Ries, 'How to Get Ideas', 1961*). Therefore, in considering how to teach and learn mathematics, it is important for both teachers and students to understand that willingness and self-help on the learners' part is very crucial. **This means that students should develop open minds and interest in their own learning**. It is only then will their subconscious pour into their minds the mathematical skills and knowledge.

John C. Maxwell argues that; *"All the good advice in this world won't help if you don't have a teachable spirit."*

(*https://johnmaxwell.com/how-do-i-maintain-a-teachable-attitude/* ,*2011*) In this case, helping students to be active, teachable, receptive, curious, and eager to and about learning of mathematics is the teachers' first task. This is how mathematics should be taught. One way to achieve this is by teachers being everything to the students – from facilitators to leaders, managers, motivators, and to being parents.

I conducted an online research about countries with very good performance in mathematics and science, and from the list compiled by Program for International Student Assessment (PISA), it was found that Singapore, China, and Japan were top on the list. How mathematics is taught in Singapore for example, and why is it different from other countries? Below is a paraphrased extract as to why Singapore has been first on the list for a number of years as told by Nicole Gorman. (*https://educationworld.com/a_news/why-do-singapore-students-surpass-rest-world-mat-and-science-1384216695*,*2016*)

The Trends in International Mathematics and Science Study (TIMSS)

This is an international assessment of sixty participating countries that proved in 2016 that Singapore's students dominated in both mathematics and science in every tested grade level because it did the following four things differently compared to other countries, which may help to raise student achievement:

1. *Foundational learning/deep mastery*

   Expert opinion was that Singaporean students were successful in mathematics because their curriculum teaches them a deep mastery of the subject through carefully calculated foundational learning where each grade level is a building block. Mathematics in Singapore focuses on children not just learning the subject for the sake of the exams, but also truly <u>mastering</u> a limited number of concepts each school year. The goal is for children to perform well because they get to understand the material on a deeper level. They don't just learn it for the test.

2. *A culture of growth mindset*

   Singapore's Ministry of Education heavily believes in "research-proven pedagogical (academic/educational) approaches that lead to lasting learning beyond the test. One of these approaches is in helping the students to obtain a <u>growth mindset</u>, which ultimately helps them to persevere, especially when they are confronted with the difficult material associated with advanced mathematics.

3. *Emphasis on visual learning*

   Emphasis is placed on model drawing, which uses units to visually represent a word problem. Students learn to visualize what a word problem is saying so they can understand the meaning, and thus learn how to solve the problem. It is in word problems where students often struggle the most when being tested on difficult mathematical concepts, but Singapore mathematics students routinely tackle them with ease due to their process of learning. Instead of focusing on the concrete meaning of the words within the problems, students <u>turn words into pictorial models</u> that transform words into recognizable pictures for the young minds. They use the model drawing approach to help the children to get past the words by visualizing and illustrating word problems with simple diagrams. And as students become better and more confident as problem solvers, they become more interested in mathematics.

4. **Mental mathematics as a core principle**

Singaporean students are encouraged to become successful at doing mathematics in their heads. Mental mathematics is one of the cornerstones of Singaporean mathematics as its emphasis is on helping them to calculate mathematically in their heads, thus developing a number sense and place value. This helps the students to not only get mathematics questions right, but get them right quickly without the interference of outside tools.

Interesting, isn't it? Should we all go the Singaporean way? It would be unwise to merely emulate the educational practices of all well performing countries such as Singapore, China or Japan. This would be against the countries' cultural and educational values. However, we can learn from such knowledge and practices, and embed them into our own education system as well as societal context. For example, instead of teaching money saving or banking using some complex bank methods, mathematics teachers can start by bringing in the ideas of the Village Banking methods. This is what most of the students have heard about, at least from either parents or friends. Some even help their parents with calculations of the very banking systems. At times teachers of mathematics will do well to invite some parents so as to allow them to help on how to teach certain concepts culturally.

Through life experiences in and outside of school, students begin to make discoveries and developing of their own ideas. I see this as the growth process. Therefore, it would be very unfair and challenging at the same time to the students if teachers are to impose their mathematical ideas and knowledge on them without considering their reasoning and contributions. Therefore, rather than teaching by just imposing mathematical concepts on students like their heads are wholly empty, learning of mathematics should be a buildup process of students' ideas.

From the education system of Singapore, we can also learn that growth of mindsets is one key factor in teaching and learning mathematics the best ways. Mathematics as a subject must be taught in order for students to think, act, and live as mathematicians and not just to pass exams. With hard work, success can be achieved. Furthermore, it must be enshrined in the teachers' and students' minds that teaching and learning of mathematics should go beyond passing of exams, and also that the school curricula should aim at preparing students for life in society where mathematical skills are inevitably needed.

Many countries, Zambia included, have remained underdeveloped for so many years because teachers and students of mathematics and science have hardly gone beyond the passing of exams. Even worse is that students are often times, restricted to too much theory intended curricula. This should be changed. There is more to learning than just passing of examinations. As a teacher, you should always relate mathematics concepts taught especially at pre-tertiary level to real life. If you can't, then do not waste the students' time teaching them. The mathematical knowledge and skills acquired in schools by students should become part of their daily life activities.

Further, it is important to understand that a good teacher is not "Mr. Know-it-all", but rather someone who understands that students – like other human beings – develop ideas, and that they are able to make great mathematical discoveries. Hence, teachers must concentrate and use correctly those ideas in building the students' understanding of the subject.

We are all ignorant about the things we have no ideas about – not that we are not intelligent, but because we have not put our ideas into it, and also because we have not had the chance to learn about them. How can students be taught if their ideas are detached from the learning process? How can they up take chances if they are not availed to them? What we know encourages us to learn more about new things, and that is the reason why teachers should build lessons from the known to the unknown.

I managed to talk to some parents and teachers of mathematics, whom I asked about their views on why some students fail mathematics, and also on what they thought was the key to the best ways of teaching and learning of mathematics. This is what they said:

Mr. Twenda, G. E of Tibi Primary School said, *"We look at maths like any other subject. We fail because mathematics is not like any other subject. Interest is key to all learning. If it is not, there can never be success."* Below are some of the points he highlighted:

- Many students don't practice regularly. Mathematics is a practical subject, except we don't do laboratory/chemical practical.
- Mathematics has its own special language which is different from the common spoken language. For instance, the word '*functions*' in mathematical context has a different meaning to that used in common spoken language. Some teachers and students ignore the language aspect.
- Some pupils have been defeated because of believing in the wrong myths like mathematics is a boys' subject.
- Background also matters. How one was taught whether it was practical or theoretical has a bearing based on having a poor background.

- Connectivity of concepts between the previously learnt concepts and the current is another important key factor.
- Parents' encouragement – a bit of a push is very helpful to the learners. Some parents are not fully engaged in their roles.

Ms. Nanfukwe, M of Nsama Day Secondary School explains; *"Many students fail because they lack interest. With interest, mathematics is simple."*

Mr. Sakeni, J who is a civil engineer had this to say: *What helps in learning mathematics is:*

- Mastering the basic principles of each topic. For example, mastering the operations
  (+-×÷) of integers will help one learn higher principles;
- Meditation of mathematical manipulations;
- Relating mathematics to real life problems and understanding how it is applied.

The students' ideas are very important in their learning. Teachers should strive to trigger much of such ideas through giving of assignments, projects, and other mathematical problems. This will help students to become not only mathematical problem solvers, but fully confident developed individuals with proper and disciplined understanding of life.

4.1 How Best Should Mathematics Be Learned?

You might have heard about this famous saying before, and yes it remains true even now – "practice makes perfect." As students of mathematics, do not be a spectators or passersby in the learning process, not even when it is via eLearning. Do not just be idle and wait watchfully for knowledge to come your way. Be active and make efforts toward acquiring of those mathematical skills you would like to gain. A good teacher encourages students in that line. Unlike with other subjects, one cannot afford to be a passerby and expect to succeed at the learning of mathematics. Its success calls for practice, practice, and more practice. There is no substitute for that, and there is no one who can do the practicing on the student's behalf.

How then can teachers assist students not to be passersby, but rather active in their learning? What I teach to my students is what I have taught myself, and always must continue doing so. A good teacher is one who remains a learner all his/her life. The best teacher of mathematics is not only a teacher to his/her students, but most importantly, to oneself. A good teacher knows how much he/she needs to improve even when is regarded as an expert by others or students.

"A teacher needs to have a bucket of water before he is able to give students a bowl of water" – Chinese proverb.

Teachers first need to teach themselves before they can consider teaching others. This is why, especially for teachers, teaching should not be a mere task or a job. Whatever might have been the reasons that resulted into one joining the teaching profession, know this: it is a very good call. It is about giving and sharing of knowledge to others. I actually consider it an honour because it is about giving light and direction in people's lives and this makes teaching a very special privilege which all teachers should be happy about. Teachers should be joyous and delightful about this great gift, and demonstrate their happiness through teaching mathematics the best ways.

In mathematics education, the training that teachers undergo does not perfect their mathematical problem solving skills. There is always a need to improve, practice, and revise their work over and over. In the world of today with new mathematical discoveries almost every day, a good teacher should always keep abreast with new developments. This quest for additional knowledge is key in perfecting the teachers' mathematical skills. This implies that teachers should effectively and continuously do their self-given tasks before they can consider giving one to the students. Their task is that of familiarizing themselves with the mathematical content they wish to offer to the students.

Teachers must in all cases be the leading examples to their students. Therefore, it is important that teachers develop and keep the right pace and attitude towards attending of workshops, seminars, church meetings, and other knowledge building programmes. In all this, students' learning of mathematics should be the center of interest.

4.2 Getting Involved

It is interesting to note that as humans, students hardly forget the things and activities to which they were a part of. Indeed, who forgets a fight in which he/she participated? In the same vein, mathematical findings and activities to which students were part of linger strongly in their minds. Students should be helped to understand that they, too, have a role to play in their own learning, and they should assume that responsibility.

Nicole Gorman, the senior Education World Contributor said; "…students are more likely to succeed particularly in difficult subjects like math and science when they ditch the feeling that they are simply not good at it, and replace it with the feeling that they can ultimately succeed if they keep trying." It is impossible to try anything when you are watching from afar. Get to the center of learning activities; get involved.

Students can take up responsibility for their own learning by having a positive contribution to the learning process. It should be understood that interest is the driving factor in such contributions. In ensuring that interest is maintained, teachers should always try to help the students to understand the relationships of mathematical concepts to daily life activities. Learning methods that trigger learners' interest and confidence such as problem solving, visual learning, deep mastery, mental principle, and discussions, among others, should be frequently used.

As students interact and consult with their parents, teachers, and peers, the interest and confidence in dealing with mathematical problems come naturally. Important to note is that it is only when a student is exposed to either peers or teachers that his/her Interest and confidence is tested, and when dealing with a problem at hand, he/she demonstrates the level of that very confidence. In as much as students' groups work as well as teacher centered approaches emphasized in learning, individual efforts and workings should also be stressed in the learning of mathematics. When students gain knowledge through individual efforts, teachers should support them. They should help them to build on that knowledge and solidify it. This does not only build the students' confidence and interest, but when it's done successfully, the teachers' efforts can be likened to the brushing of a shoe which has been briefly polished. The shoe (student) just fully shines because there is already polish (brief knowledge), which makes the brushing easier and effective.

In fact, a good teacher is mainly a facilitator in the learning process, and understands the importance of appropriate teaching strategies such as deductive and inductive reasoning, and uses them when suitable. Due to some people's myths and prejudices about mathematics, many people have been engulfed in fears of the challenges that are faced when dealing with mathematics learning, and to some, that has resulted in them hating mathematics in general. The hate is normal, but that's not the solution!

A mathematical approach is key to many of the world's problems today, and when we talk of fear for mathematics, it is important to note that there is no better remedy for fear than when love has been found. Love is attractive. A student of mathematics must develop love for mathematics and the fears to its challenges will disappear. Once this love or interest is established, students will be able to conquer their fears and face their challenges in learning mathematics the best ways.

I did my senior secondary education at Butondo Secondary School in Mufulira. During my school days, I was privileged to have had the opportunity to learn from one of the greatest mathematicians Zambia has ever produced, Professor Frank Tailoka. Prof. Tailoka is a former Copperbelt University (CBU) professor of Mathematics and Decision Science. At the time of authoring this book, he served as the Examination Council of Zambia (ECZ) Board chairperson as well as Mukuba University director. He, too, is actually a former student of Butondo Secondary School.

In an effort to encourage and motivate students to reach greater heights particularly in mathematics and science, Prof. Tailoka finds time to visit his former secondary school. When he does, he usually encourages students to work extra hard and organizes awards for the top performers in the very subject, mathematics. Back in the years as a secondary school student, I attended some of those award giving ceremonies, and whilst there, Prof. Tailoka said something that permanently changed my perception towards learning.

He briefly shared his secret to success in life as a student of mathematics and a teacher. He said; *"My secret to success is that I got married to my two wives."* As a student, I was confused in that moment. I mean, this is not what I had expected to hear as the secret to academic success, especially for mathematics education. But he continued and said;

"I love my first wife more than I do the second one. I love her so much not because I have known her since childhood, but because unlike my second wife, I know for sure that she cannot leave me. No matter the situation, I believe this woman will always be by my side. She is all loving, and I shall always dwell in her love even to my grave. I know even some teachers here are wondering about what I am talking about, knowing that I only have one human wife. I am sorry to disappoint you, but I have two wives. The first wife, ladies and gentlemen, are the books. Since childhood, I have been in constant contact with books and the knowledge (love) I have gained in return never disappoints. Nobody has the power to get it away from me, not even my parents who sent me to school. The knowledge you get from books will forever be with you to help you survive in society."

Students must always strive to be in love with books. The love or knowledge in books never disappoints. There are a lot of very good mathematics books for both content and methodology. Your reading this book is an indication that you are in the right direction. Do not stop here; continue searching for more knowledge. Mathematical knowledge helps one to become a problem solver. This is what this world is in dire need of – problem solvers. Therefore, enabling students to develop the love for mathematics is another great task for teachers. This involves offering of material support such as books and mathematical equipment. It can also involve creating a good relationship between teachers, parents, and the students. Furthermore, it involves knowing students well and by name, and also to understand their specific learning weaknesses and strengths. It calls for employing of both individualistic and non-individualistic approach to the students' learning.

In the teaching and learning of mathematics, it should be understood that it is not a one man task. Teachers, students, and parents should not work in isolation. There should be unity in their workings, and once this partnership is successfully established, people's myths and prejudices about mathematics, the learners' fears of it, and its challenges will all evaporate, and it shall be replaced with the love for mathematics. Teachers should not over emphasize the abstract part of learning mathematics, but should stress how concepts relate to man's daily activities. For example, teachers can come up with a real life activity to demonstrate the concept of Linear Programming mathematics. At times, some people do think of science subjects such as chemistry, biology, and physics as being more life related than mathematics. But little do they realize that mathematics is the greatest science of all sciences.

It is how a subject is studied that makes it a science. The way mathematics is studied and its applicative value to man's life makes it a science. When we talk of mathematics, it at times encompasses all those subjects that in whatever sense uses the mathematical approach in its dealings and findings. Hence, when teaching mathematics, students should be made to feel worthy of learning it.. The most important step is by enabling students to feel indebted to the powers of mathematics in shaping man's life. Once this understanding is developed about the importance of the subject to man's development, then students shall develop the free will to contribute to their learning.

CHAPTER FIVE

5. STUDENTS' AND PARENTS' ROLES IN THE TEACHING AND LEARNING OF MATHEMATICS

At birth, what a child will become is unknown. Just like a seed, they are in the process of becoming something not yet known. While at home, it takes the parents' and guardians' effective guidance for any child to develop into a fully-fledged member of society. Effective parenting involves the realization that raising children is a unique privilege which should be fulfilled with joy. The scripture in *Psalms 127:3-5* reminds us that children are a gift from God. Parents understand that their responsibility towards their children's growth is not as easy. This is why they share part of that with the teachers. Teachers are the children's second parents.

5.1 Teachers' and Students' Roles

Teachers are the parents' substitutes in schools. Therefore, they, too, should fulfill their responsibilities joyously. The teacher-parent relationship is very crucial because children spend most of their time in schools than they do at home. This highlights the importance of cooperation in the learning processes between the parents, teachers, and the children.

A teacher is the manager of a class, is in-charge of all the classroom proceedings, and facilitates the individual efforts of the entire class. It is the teachers' role to ensure that the class is well managed, motivated, and that the best choices in terms of teaching methods and techniques are applied in order to maximize the learners' potential in mathematics. Therefore, teachers should know their student's strengths and weaknesses, and cultivate the most positives out of them.

Teachers must spell out clearly the school rules that students are to follow, and at times, a little bit of flexibility to the students mistakes can create a friendly learning environment. This, too, is very important in mathematics education. Also, a teacher who deals with student issues from afar can hardly succeed in helping them. Teachers instead should be able to easily engage with the students, interact with them on a more personal level, and help them to become problem solvers. They should individualize their teaching methods in order to diagnose the challenges faced by every student as well as to create a bond like the one that exists between children and their parents. Remember, children are more at ease when they are with their parents than with other people. It is the reason why they can express themselves easily in many instances.

In spite of whether or not the above stated teachers' roles are being fulfilled, students too have important roles to play in the best ways to teaching and learning of mathematics. California Department of

Education, 1999 guides: "Proficiency in mathematics is not an innate characteristic; it is achieved through persistence, effort, and practice on the part of students, and rigorous and effective instruction on the part of the teachers." Students and teachers should supplement each other's efforts – their roles are intertwined.

To begin with, students are obliged to follow the well laid down school rules and guidance provided in all disciplines. It would be detrimental for a student to think that progress in mathematics learning depends entirely on the effectiveness of the teachers' methods and activities. Most students have both great desire and fear to learn new mathematical concepts. However, they need someone who can tame their fears, maximize on their curiosities, and help them to learn mathematics joyfully and effectively. As earlier stated, a good student plays a part in his/her own learning. They are not passersby, but rather, people who consider individual search for extra knowledge.

Every student should understand that no matter how effective a teacher may be, he/she cannot teach 100% of the topics or subject. This is because of the normal and unavoidable life circumstances such as ill health, global pandemics, attending of assemblies, among others. It is because of these unavoidable circumstances that students should search for extra knowledge from other sources so that the missing of some concepts in class can be compensated for through personal efforts.

When faced with mathematical challenges, a mathematics student will do well to consult teachers, parents, friends, and/or the internet. By so doing, the student will gain a wider perspective on how to deal with mathematical challenges the best ways. Students should be free to ask wherever they lack understanding. Teachers are not saints; they, too, are just ordinary human beings. Therefore, if fellow human beings such as teachers, doctors, engineers, economists, lawyers, agriculturalists, et al, can do it – and can deal with the mathematical challenges successfully – why can't students do it as well? They, too, can do it. Why not imitate those who have excelled at it? This can be very helpful.

Students should understand that critiques from teachers, peers or parents are part of the learning process, and that they are not to be shunned. In this case, the main aim of critiquing in learning is not to expose someone's ignorance, but to enable one to grow, to think critically, analytically, and to become a great problem solver. Note: Even teachers' arguments are open to criticism, but always keep it in mind that in whatever criticism, the conclusions should be of benefit to both the students and the teachers.

As elderly people, teachers should be highly respected. They are the parents in schools; they offer the guidance to the students, and they have got the school life experiences – a thing that cannot be surpassed by an ordinary person. Just like great sports men and women do in their individual trainings, exercises, and desires to improve, so is the student of mathematics supposed to be. He/she should arrange for individual trainings of the mind to deal with mathematical challenges – do the individual revision work, and always strive to improve and excel in learning of mathematics. This amplifies the help rendered by teachers, peers, parents, and others sources.

5.2 Parents' Roles

Every child has a right to education; it is every parent's duty to educate their children. However, this responsibility does not solely lie on parents. No wonder they delegate part of their role to the teachers in schools knowing well that they share a common goal, that of nurturing the children. What then are the roles of parents in the teaching and learning of mathematics to their children?

It is the parents' duty to ensure that every child is availed the right to education. This does not just mean taking children to school, paying their school fees, and providing other financial needs. It means more than that. Parents are responsible for the provision of their children's primary basic needs. These needs include food, shelter, clothing, security, love, and belongingness. All these provisions make up for an environment conducive for learning.

The arrangement of students spending more time in school with their teachers and peers than at their homes is a great one. However, it still remains the parents' duty to ensure that the child being in school is of great benefit in as far as learning is concerned. In learning, parental support is very crucial. This support can be financial, moral, emotional, physical, and/or spiritual. It is the support of a very special kind. To fulfil this, parents do not need to be saints or geniuses or very rich. No! But because they have had the school life experiences, even just a little, that is sufficient in fulfilling their role. If the parent cannot fully offer the academic support to the child such as helping during homework, at least, he/she should have some ideas on where the children can get the best help from. Why not direct and guide them in such a way? Overall, guidance and support in every aspect of life to a child is vital.

Living in economically unstable societies where at times even the provision of fundamental human needs may be hard. But then again, to be good parents does not depend on riches. Whole support accompanied with love is what matters, and has a great impact in the child's learning of mathematics. It is the parents' duty to enable children to understand that in life, one can use past and present situations to improve the future.

Prof. Tailoka, whom I talked about earlier, once told us this: *"Looking at you in school uniforms makes me to recall my school days as a pupil. Just like you, I once was a pupil here* (at Butondo Secondary School), *learning from the same human teachers, in the same classroom blocks, and perhaps sitting on the same desks where you seat now. By the way, I grew up in Kankoyo Township."* (Kankoyo

Township is one of the devastated townships on the Copperbelt province of Zambia)

Do not let your current situations deter you. If anything, let it be the motivation to reach greater heights. Therefore, discipline, focus, determination, and hard work are the keys to success, in and out of school life. Parents, teachers, and the community should work together and strive hard to kill – and yes, I mean to kill – the myths and prejudices of some students about mathematics as being a difficult and a boys' only subject. Those assumptions are wrong.

Seriously, if mathematics was as difficult as some people think it is, I would not have gotten this strength and motivation to put my thoughts and experiences to write this book for it to bear testimony to these words, and that hopefully others would also learn from it. I for one do not consider myself to be a genius mathematician, somebody whose pass percentage in mathematics at Grade 9 (final examinations) was 45%, 19% at Grade 10 (mid-term exams), 52% at Grade 11 (mock exams), 78% at Grade 12 (mock exams), and graduated with a merit in mathematics at senior secondary (Grade 12 final examinations).

A student may, at times, perform poorly in mathematics due to many reasons, which sometimes may be life circumstances or poor academic background. Therefore, having a non-impressive academic background does not make a dull student, and neither does the opposite make a genius. Mathematical ability is a natural gift. It is a special gift that every human being is born with. It emerges at different stages of students' lives, of course through encouragement and hard work. What matters most, is the early and fast recognition and nurturing of this ability in a student, and the maximization of its potential to grow to greater heights.

> *"A growth mindset, as defined and popularized by Carol Dweck, is the idea that intelligence is not a set of fixed traits, but rather is something that can be developed and improved through hard work and education." – Nicole Gormon*

From my academic stats stated above, who would have even thought that a student with such a poor mathematical record would one day improve to such impressive levels? Who would have thought that such a student would entirely fall in love with mathematics, go further and study mathematics content at both diploma as well as degree level, and become a teacher of this very subject? In mathematics education, interest, hard work, and discipline pay off. Everyone else can learn mathematics successfully. I am forever grateful to my parents, teachers, and peers for their immeasurable support. Despite all the challenges, you joyously helped me to be what I have become today.

Students always remain indebted to their parents and guardians, whose lives are a dedication and commitment in helping them realize and reach their full potential in the process of learning.

5.3 Roles of Cooperating Partners

Just like those of students, teachers, and parents, the roles of cooperating partners in fostering the teaching and learning of mathematics cannot be ignored. The roles of the policy makers, governmental, and non-governmental organizations is of great importance. In most cases, emphasis is placed more on teachers and parents as the custodians of students and learning ideas. However, we all have a part to play as a people.

In this regard, the cooperating partners should still continue with their support through offering of grants in schools, student bursaries, and the provision of learning environments that are conducive as well as the provision of learning materials. There is still a need on their part to even further their support through tracking the great students of mathematics so that they reach the highest level – and that they do not divert into other fields or perhaps to other areas than where there knowledge is desperately needed.

CHAPTER SIX

6. HOW TO SUCCEED AT MATHEMATICS EXAMINATIONS

In the field of education, to examine is to ascertain the effectiveness of the learning processes in a particular field. Most people develop a phobia towards mathematics examinations. Fear for examinations is common, and to succeed, you don't need to punish yourself over it because it is normal. Examinations are an act of testing someone's mind, strengths, capabilities, intelligence, and the general potential they possess.

Most students do fear examinations because they understand that maybe, they might not meet the expected performance standards. This fear is good especially if it is taken positively. It can enable the student to feel motivated, work hard, and prepare more for what is expected in the exams. It is important to understand that examinations and any other type of assessment are an integral part of the learning process. Students should understand that examinations are used not as a tool to expose those who might be poor performers, but to assess the effectiveness of the learning process, the students' conceptual understanding, as well as the teachers' methodology.

To achieve success, it calls for self-awareness on the students' part about what is being done, and knowing the true value of what someone is doing (the importance of the course to you as a student and to others around you), and why it must be done successfully. This will enable one to set goals and work towards reaching them.

6.1 Ten Points on the Best Way to Prepare For Examinations

Steps: Having adequate preparation for examinations is an important part of learning, and below is how a good student must do it:

- **Start now;** practicing, studying or revising should all start right from the beginning of a course. It would be detrimental for a students to waste their first, second or third years of study and think of only doing it when exanimations are near or due. Remember, time wasted is never recovered. Preparations should always start as soon as the school term, semester, and year of study begin, and should continue right through until the exams are due. This is so because the earlier one starts moving continuously, the greater the distance covered in making those preparations a success.

- Before facing the official examinations, it is always important for a student to **test and examine oneself**. However, whether or not he/she passes the personal exams, there is no need to get one's head down or get too excited. Instead, he/she should happily take it as a motivation to do better next time or to maintain the best performance. There is no need to feel too satisfied or get too relaxed for a positive achievement, but should feel positively motivated to do better in the future, strive to beat previous records, and seriously make it the reason for improvement.

- **Revise regularly;** instead of studying mathematics for two, three or four days only in a week, make mathematics study your daily routine. Let there be other subjects on your time table if possible, but whatever subject(s) is on the time table, you should always start or end by revising at least three mathematics problems on a daily basis. That makes at least 21 solved problems in a week, and above 80 in a month. You can read that again! Exams will be simple.

- **Be a good time manager;** make a good study plan. Find time to relax, have leisure time as this keeps the mind always strong to deal with mathematics challenges with ease.

- **Get familiar with exam type of questions;** as a student, the more acquainted or friendly you become with exam type of questions, the better for you in answering them the best ways. This is very important in building self-confidence towards problem solving.

- When presented with an examination task or paper**, do not rush into answering.** Students make a lot of mistakes and errors when rushed into answering mathematics problems. Therefore, it is always wise to wait a second, to let your body, mind, heart, and soul to get connected. Give yourself a second to think about what you are about to do, and how important it is to your life, and that of others around you. Take a very deep breath and pray to God that you may be helped to remember what you have studied, and then you can start answering. You will be amazed at the results.

- **Be strategic;** in dealing with examination tasks. Always begin with that which you can easily handle or remember, and make sure that you do it correctly, then that which you have some shortcomings with should follow later. This helps in keeping the mind strong for the final challenges.

- A student should understand the **importance of the course of study**. By understanding, it means being aware of the personal importance attached to the aims and objectives in a particular field, and what contributions can be made in achieving them.

- No matter how competent or incompetent one is, a student should not study in isolation. Students should be ready to contribute and share ideas with their peers. We all **learn from others**, implying that there is a need to stimulate one another as a way of preparing for exams.

- When it is finally time for the examinations, do not panic. Instead, **be content** with what you already know. Instead of panicking in the last minute in search for new knowledge and ideas that cannot be fully consolidated, understood or revised due to limited study time, it is better to revise, understand, and consolidate on the already acquired/learned knowledge – at least where one has brief knowledge. This mostly is the case with some students during examinations time. It does help to concentrate on few things especially when time is limited. Save that last minute for brain relaxation. This is quite important.

CONCLUSION

YOU have read all the content of this book, right? Very good! Now, what is next for you as a student or a teacher of mathematics? What is next for you as a parent?

To students, I know for sure how much you wish to have and maintain improved grades in mathematics. To the teachers of this subject, I am also aware of your desire to make the teaching and learning of mathematics interesting and effective. To fellow parents, your desire for your children to acquire the best grades in mathematics is unquestionable. You all can have this. Below is a model guide to teaching and learning mathematics the best ways.

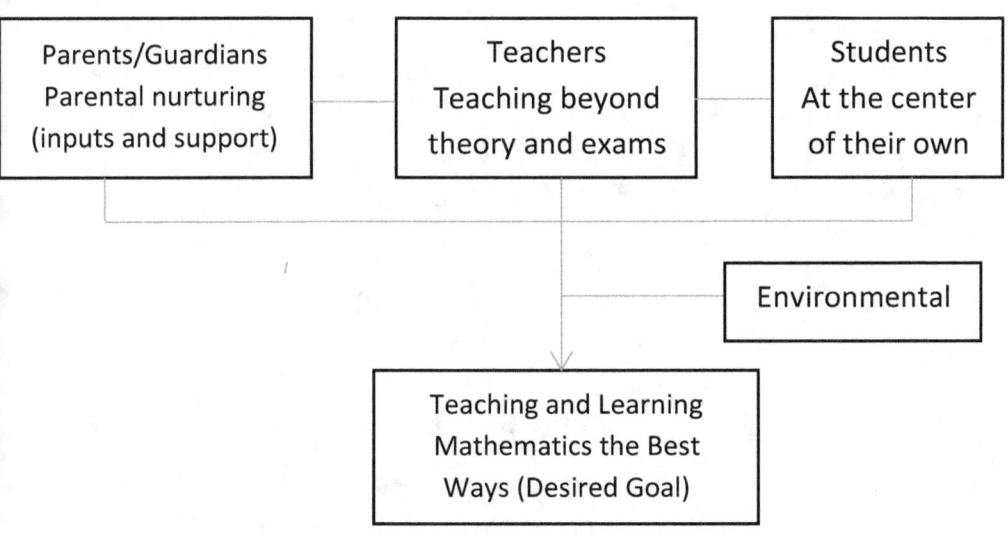

Figure 3: model guide to teaching and learning mathematics

It is my sincere hope that you have learnt a thing or two about 'Teaching and Learning of Mathematics the Best Ways'. I would want you to keep this in mind: that which is not put into use or practice is of no value. Therefore, if the guidance offered in this book is to be of any value or meaning to YOU, I implore you to seriously put into use all that have learned. You must strive to fulfill your roles in teaching and learning mathematics beyond theory and passing of exams, and prominence at mathematics education shall be achieved!

www.ingramcontent.com/pod-product-compliance
Lightning Source LLC
Chambersburg PA
CBHW080524220526
45465CB00006B/2585